Nitrogen Deficiency in Sitka Spruce Plantations

ISBN 0 11 710290 3
ODC 424.7 : 181.3 : 174.7 Picea sitchensis : (410)

KEYWORDS: Nutrients, Forestry

Enquiries relating to this publication
should be addressed to:
the Technical Publications Officer,
Forestry Commission, Forest Research Station,
Alice Holt Lodge, Wrecclesham,
Farnham, Surrey GU10 4LH.

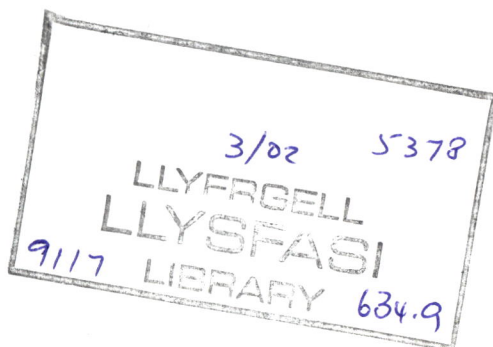

Front cover: 14-year-old Sitka spruce on an
upland raised bog overlying Ordovician, show-
ing a good response to nitrogen fertiliser applied
3 years before. (*C. M. A. Taylor*)

Inset. Sitka spruce in severe growth check from
nitrogen deficiency. (*C. P. Quine*)

Contents

Nitrogen Deficiency in Sitka Spruce Plantations

Summary

On moorland and heathland soils in Great Britain nitrogen deficiency can severely restrict the growth of certain conifer species, including Sitka spruce, the main commercial species. Until the 1970s this was thought to be due solely to competition from heather and was commonly known as 'heather check'. However, increased planting of Sitka spruce on very nutrient-poor soils revealed that, even after removal of heather by herbicide treatment, growth was still limited by low availability of nitrogen. This can be caused by limited soil nitrogen capital and/or slow rate of nitrogen mineralisation. Application of nitrogen fertiliser can overcome this deficiency although several applications may be required to achieve full canopy closure. Once this stage is reached demand for nutrients is reduced due to shading of competing vegetation, improved nutrient cycling and capture of atmospheric nutrients and further inputs of nitrogen should not be required.

The major difficulty facing forest managers in determining the treatment of a nitrogen deficient stand is deciding whether heather control, application of nitrogen fertiliser, or a combination of both, will yield the most cost-effective response on any given site. This Bulletin explains the background to the problem, categorises the range of site types involved, and advises on the treatment available.

La Carence d'Azote dans Les Cultures de l'Épicéa Sitka

Sommaire

Sur les sois de bruyère ou de lande dans Le Royaume-Uni, la carence d'azote peut limiter la croissance des résineux y compris l'épicéa Sitka, l'espèce principale commerciale. Jusque les années 1970 on a pensé que ce phénomène était imputable seulement à la concurrence par la bruyère, ainsi le terme 'arrêt-bruyère'. Cependant, avec l'augmentation des plantations de l'épicéa Sitka sur les sols avec contenu très faible en substances nutritives, on a remarqué que même après l'éloignement de la bruyère par des herbicides, la croissance était encore limitée par la disponibilité faible d'azote. La raison est le contenu faible d'azote dans le sol et/ou la vitesse faible de minéralisation d'azote. L'application des engrais azotes peut vaincre cette carence, bien que quelques applications soient nécessaires pour accomplir couvert plein des arbres. Quand la culture atteint cette phase, le besoin en matière nutritive est réduit par l'ombragement de la végétation de concurrence, la circulation meilleure des éléments nutritives, et la capture des éléments nutritives de l'atmosphère; il ne faut pas des autres applications d'azote.

En determiner le traitement d'un peuplement avec défaut d'azote, la difficulté principale pour les administrateurs forestiers est décider si la lutte contre la bruyère, l'application des engrais azotes, ou une combinaison des deux fournira le résultat le plus efficace et économique sur une station donnée. Ce Bulletin explique le fond de ce problème, catégorise les types des stations affectées, et donne renseignment sur les traitements disponsibles.

Stickstoffmangel in Sitkafichtenkulturen

Zusammenfassung

Auf Moor- und Heideböden in Grossbritannien kann Stickstoffmangel das Wachstum gewisser Nadelbaumarten einschliesslich der Sitkafichte (der Brotbaum) stark einschränken. Bis in die 1970er Jahren hat man gedacht, dass die Ursache dafür nur auf Heidekrautkonkurrenz zurückzuführen war, und daher war die Erscheinung als 'Heidekrautstocken' bekannt. Die zunehmende Bepflanzung von Sitkafichte auf sehr nährstoffarmen Böden hat jedoch gezeigt, dass das Wachstum sogar nach Heidekrautvernichtung mit Herbiziden noch durch niedrige Stickstoffverfügbarkeit beschränkt war. Die Ursache davon kann durch beschränkten Stickstoffinhalt im Boden und/oder durch langsame Stickstoffmineralisierung verursacht werden. Die Anwendung von Stickstoffdüngemitteln kann diesen Mangel überwinden, aber wiederholte Behandlungen mögen nötig sein, um vollen Kronenschluss zu erreichen. Einmal wird dieses Stadium erreicht, so wird der Nährstoffbedarf durch Beschattung von den konkurrierenden Pflanzen, durch verbesserten Nährstoffkreislauf und durch Nährstoffeinfangen aus der Luft vermindert; weitere Stickstoffbeigaben sollen nicht nötig sein.

Die Hauptschwierigkeit für Forstverwalter bei der Behandlung eines stickstoffbedürftigen Bestandes besteht darin zu entschliessen, ob Heidekrautbekämpfung oder Stickstoffdüngemittelverwendung oder eine Kombination der beiden auf gegebenem Standort das wirksamste und wirtschaftlichste Ergebnis liefern wird. Dieses Bulletin erklärt den Hintergrund dieses Problems, bestimmt den Bereich der einschlägigen Standortstypen, und gibt Beratung über die Behandlungen, die zur Verfügung stehen.

Nitrogen Deficiency in Sitka Spruce Plantations

C. M. A. Taylor and P. M. Tabbush, Silviculturists, Forestry Commission

Background

In the 1930s, 40s and 50s the poor growth of young Sitka spruce (*Picea sitchensis* (Bong.) Carr) plantations suffering from nitrogen deficiency was associated with competition from heather (*Calluna vulgaris* (L.) Hull), mainly on heathland soils (Weatherell, 1953) but also on moorlands (Fraser, 1933). This was thought to be a straightforward competition for nutrients and moisture (Leyton, 1955), although Laing (1932) had previously conducted experiments which indicated that decaying heather leaves in water culture produced toxic effects on the root system of Norway spruce (*Picea abies* (L.) Karst.) seedlings. Before the advent of suitable herbicides early attempts were made to suppress the heather physically by mulching (Zehetmayr, 1960) and although this stimulated growth (Leyton, 1954) it was never proposed as a practical treatment due to the high input of labour.

It was found in Forestry Commission experiments that admixture with Scots pine (*Pinus sylvestris* L.), lodgepole pine (*Pinus contorta* Dougl. var. *contorta*) or Japanese larch (*Larix kaempferi* (Lamb.) Carr.) improved the growth of Sitka spruce due to the ability of these species to suppress the heather (Zehetmayr, 1960). Large areas of such mixtures were established on the heathlands in the 1950s and 1960s, although the poorest sites were still planted with pure pine.

On the moorlands planting of Sitka spruce was mainly confined to the more fertile, flushed bogs and peaty gleys dominated by purple moor grass (*Molinia caerulea* (L.) Moench). Growth check was recognised to be a problem on the *Calluna/Trichophorum* peaty gleys and the unflushed bogs and at that time these sites were generally regarded as unsuitable for Sitka spruce and either left unplanted, or planted with pure lodgepole pine. Although heather was present on many of these sites, severe phosphate deficiency and waterlogging also appeared to be major growth-limiting factors (Zehetmayr, 1954).

In the 1960s and 1970s the development of improved cultivation, fertilisation and weed control techniques encouraged the planting of pure Sitka spruce on heaths and unflushed moorlands, although lodgepole pine was still the favoured species on the most infertile peats. During this period Handley (1963) concluded that heather 'check' was caused by allelopathic substances produced from the root system of the heather, inhibiting mycorrhizal development in the spruce as suggested earlier by Laing (1932); although this was not fully accepted until confirmed some time later by Malcolm (1975) and by Read (1984). It also became apparent that, on certain site types elimination of heather alone would not be sufficient and inputs of nitrogen fertiliser would also be necessary (Mackenzie, 1974). This was particularly apparent on the unflushed peats where nitrogen deficiency was found to occur even when heather was not present on the site, due presumably to a low mineralisation rate (McIntosh, 1983). The short growth response period of 3–4 years following an application of nitrogen fertiliser was also identified (Dickson and Savill, 1974). These additional costs were accepted as it enabled pure Sitka spruce to be grown instead of lower yielding pine and avoided the perceived problems of managing mixed stands.

As a result, large areas of pure Sitka spruce in private and State forests are being treated by application of nitrogen (2–3000 ha yr^{-1}) to maintain satisfactory growth rates. This programme is only carried out in plantations that have not achieved full canopy closure. Once this stage is reached, demand for nutrients is reduced due to shading of competing vegetation, improved nutrient cycling and capture of atmospheric nutrients (Miller, 1981) and further inputs of nitrogen should not be required. Certainly, there has been no consistent or predictable benefit from applying nitrogen to pole-stage Sitka spruce stands in this country (McIntosh, 1984).

Subsequent increases in the cost of nitrogen fertiliser, logistical difficulties in completing large heather control programmes with herbicide due to poor weather conditions and the difficulties of aerial fertiliser applications in crops with scattered areas of checked growth, forced a re-assessment of species choice. Many sites are now being planted with species mixtures – nursing Sitka spruce with Scots pine on heathlands and with lodgepole pine on unflushed moorlands (Taylor, 1985). These mixtures provide a robust silvicultural practice which does not incur the considerable expenditure on remedial treatment required to prevent pure stands of Sitka spruce declining into check on such sites.

Recent work has demonstrated that the nurse species not only suppress competing heather but also have the ability to increase the amount of nitrogen available to the spruce, although the processes involved are not fully understood (Miller et al., 1986). Several factors appear to be involved of which improved nutrient cycling, the 'dilution' effect of the nurse species and mycorrhizal associations may be the most important.

Recognition of nitrogen deficiency

Visual identification
Nitrogen is an essential constituent of many important plant substances, including proteins, nucleic acids and chlorophyll, and thus is vital to metabolic processes determining the production of new tissue. Nitrogen deficiency has very serious implications for volume production.

The first symptom of nitrogen deficiency in Sitka spruce is a general lightening in the colour of the foliage from the normal dark blue-green colour (Binns et al., 1980). This will occur over the whole tree, and in severe deficiency the foliage becomes yellow-green, or even yellow. Subsequently, new needles will be shorter, the leading shoot becomes spindly and height growth declines rapidly. Due to the one year delay between colour change and effect on height growth, regular annual inspections of susceptible areas in October and November (before the onset of any winter damage – Binns et al., 1980) are advised to ensure that deficiency is promptly diagnosed and treated before height growth is seriously impaired.

Foliar concentrations
The other method of diagnosis is to measure the foliar nitrogen concentration. Although there are some doubts over the value of this technique in older crops, due to nutrient cycling within the stand, it is still suitable in the majority of crops where nitrogen deficiency is likely to occur, i.e. prior to canopy closure (McIntosh, 1983). Using the approved method of sampling (Everard, 1973) subsequent results from foliage analysis can be compared with the deficiency and optimal levels set by Binns et al. (1980) and confirmed by Taylor (1987), which are detailed in Table 1.

Foliage sampling and analysis should be necessary in difficult cases where doubts exist following visual inspection.

Table 1. Deficient and optimal foliar nitrogen concentrations for Sitka spruce (<3.5 m tall), as per cent oven-dry weight

Optimal level	>1.5
Marginal level	1.2–1.5
Deficient level	<1.2
Severely-checked growth	<1.0

Site categories
Forest managers often find difficulty in understanding or anticipating the growth response to remedial treatment. To illustrate how growth responses to heather control or nitrogen fertiliser vary across different site types, some recent experimental data on the effects of different rates of nitrogen fertiliser and control of heather on the growth of Sitka spruce are presented in

Figure 1 (from Taylor, 1987). These sites (Figure 2) have a range of soil types and lithologies which are listed in Table 2, together with the relative growth response to the treatments over a 6-year period. At Clydesdale and Kershope (peat sites) the growth response to heather control was as large or larger than the response to application of nitrogen. At Strathardle and Inverliever (mineral and organo-mineral soils), the response to the application of nitrogen was larger than to heather control, with both responses being larger than at the other two sites.

Therefore a categorisation of nitrogen deficient sites is needed to aid decisions on treatment prescription. Based on experimental data and field observation, it is possible to distinguish four categories of nitrogen availability, which are described below. These are also diagramma-tically represented in Figure 3 which indicates, for each category, the effects of a range of treatments on the height increment of Sitka spruce.

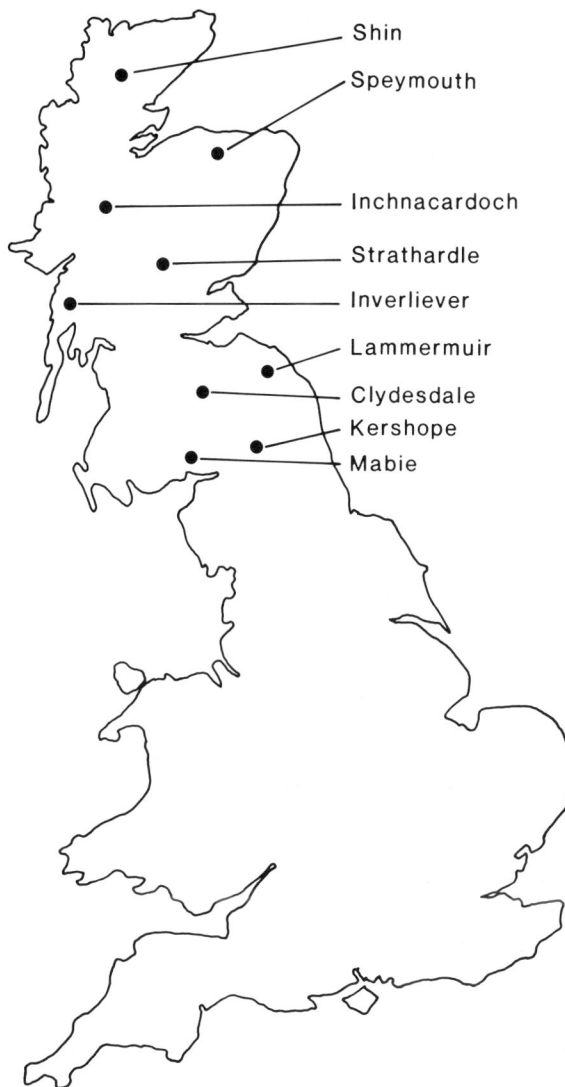

Figure 1. *Effect of rate of application of nitrogen (N) and heather control (H) on 6 year height increment (expressed as a percentage of control (O)) of Sitka spruce on four experiment sites.*

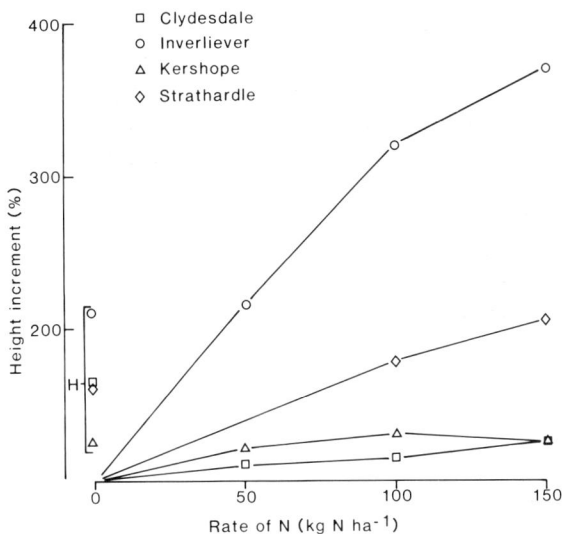

Figure 2. *Location of experiments.*

Table 2. Soil types and lithologies for the experiment sites, and the relative growth responses to heather control (H) or application of nitrogen (N)

Experiment	Soil type	Lithology	Growth response
Clydesdale	Upland raised bog	Carboniferous	H>N
Kershope	Unflushed blanket bog	Carboniferous	H≥N
Strathardle	Intergrade ironpan soil	Dalradian mica schist	N≥H
Inverliever	Peaty gley	Quartzite	N>H

3

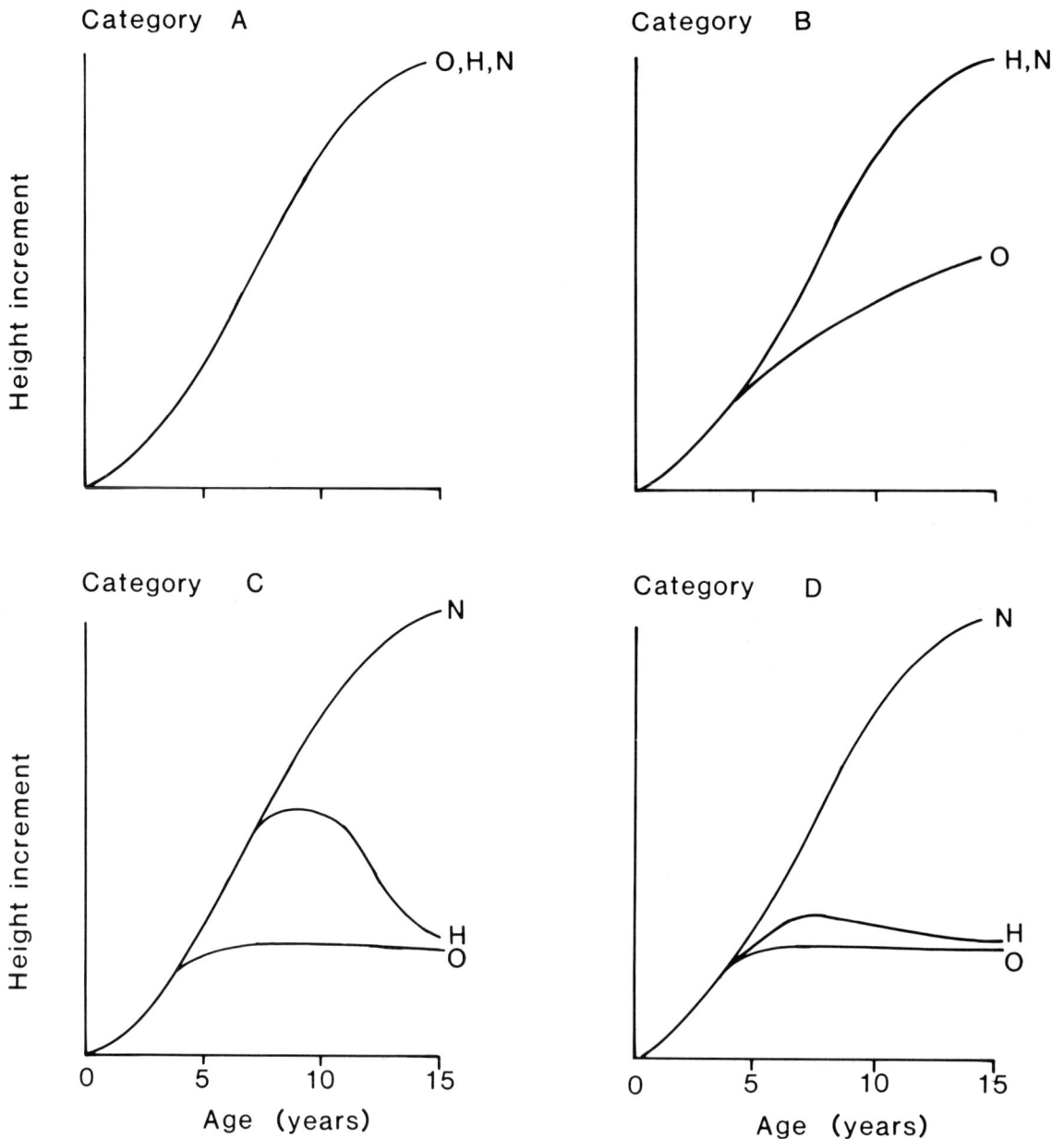

Figure 3. *Diagrammatic representation of the effects of treatments O (control), H (heather control) and N (regular application of nitrogen) on the height increment of Sitka spruce.*

Category A

Sites where there is sufficient nitrogen available for acceptable tree growth, despite the presence of heather. The inhibitory effect of heather seems to be reduced when soils are rich in available nitrogen, and Sitka spruce is unlikely to suffer any real check to growth, although there may be a slight yellowing of foliage in the 2 or 3 years prior to canopy closure. No herbicide or fertiliser is required (see Plate 1).

Normally these are sites where the heather is mixed with fine grasses, such as *Agrostis, Festuca* and *Anthoxanthum* spp., in the transition zone between grassland and heathland; or weakly flushed moorland sites dominated by *Myrica* and vigorous *Molinia*; or sites heavily colonised by broom (*Cytisus scoparius* (L.) Link) or gorse (*Ulex europaeus* L.).

Category B

The sites in this category are those where heather is the principal cause of nitrogen deficiency and successful heather control would result in adequate availability of nitrogen for Sitka spruce (see Plate 2).

These are usually heathlands on more fertile lithologies (e.g. basic igneous, phyllites, pelitic schists) or western *Molinia/Eriophorum* moorlands where the heather is sub-dominant.

Category C

Heather is the dominant type of vegetation on these sites, but is not the sole cause of nitrogen deficiency. The low mineralisation rate is also a major factor and although heather control will result in a cost-effective growth response, it will not bring permanent relief from nitrogen deficiency and subsequent inputs of nitrogen fertiliser will be required to achieve full canopy closure (see Plate 3). This category can include moorland sites where *Molinia* and *Trichophorum* are co-dominant with *Calluna* and certain heathland soils with low organic matter content.

Category D

The principal cause of nitrogen deficiency on these sites is the low mineralisation rate. Heather control will not give a cost-effective growth response. In fact, on many of these sites heather is either not present or very sparse. Several inputs of nitrogen fertiliser will be required to maintain a reasonable growth rate in Sitka spruce and enable the crop to achieve full canopy closure (see Plate 4).

Normally this category will contain lowland and upland raised bogs but it also includes podzolic soils with low organic matter on quartzitic drifts.

Experimental evidence

Experimental examples of categories B, C and D are presented in Figures 4, 6 and 5 (see also Figure 2 for location).

In Figure 4 annual height increment is plotted for various treatments in an experiment at Lammermuir in the eastern Southern Uplands on a heather-dominated peaty gley over Lower Old Red Sandstone. Here herbicide treatment (H) and nitrogen application (N) are compared with a control treatment (O), over the 6–12 year age period. In September of year 6 the herbicide, glyphosate, was applied at a rate of 1.5 kg acid equivalent (a.e.) ha^{-1} (4.2 litres product ha^{-1}) achieving a 92% kill of the heather. Then in May of years 7 and 10 nitrogen fertiliser (urea) was applied at 150 kg N ha^{-1}. The growth responses to these two treatments follow a similar pattern indicating that this is a category B site (see Figure 3). Growth in the control treatment has been less, although it has not declined into severe check, indicating that

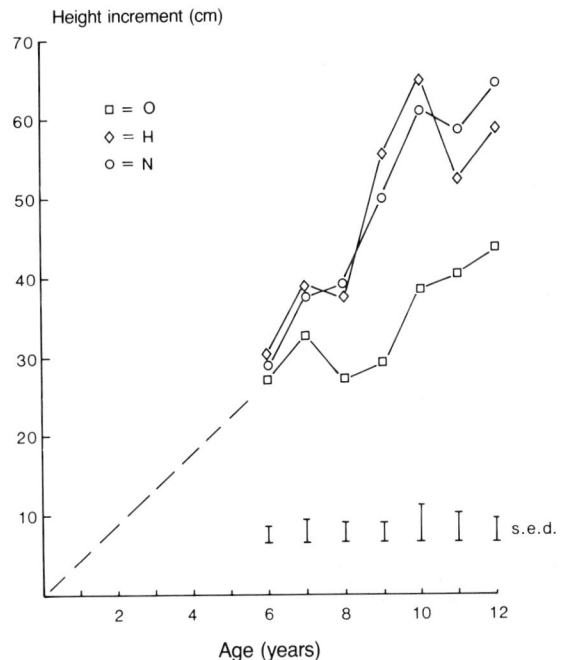

Figure 4. *The effects of treatments O (control), H (heather control in year 6) and N (application of nitrogen in years 7 and 10) on the annual height growth of Sitka spruce on a heather-dominated peaty gley at Lammermuir.*

5

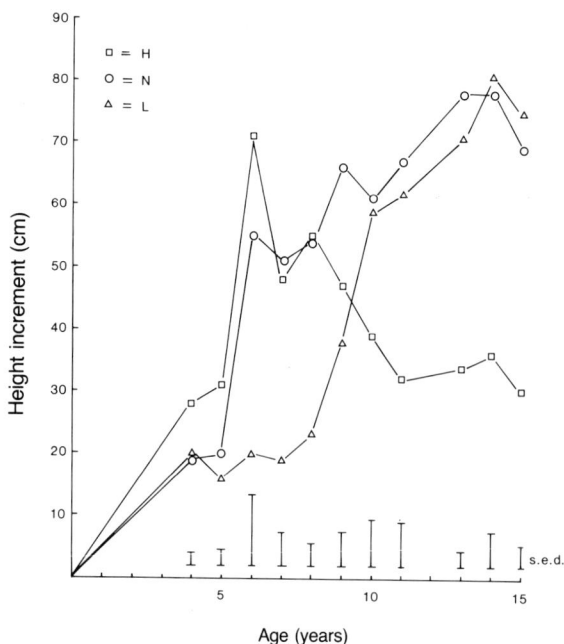

Figure 5. *The effects of treatments H (heather control in year 3), N (application of nitrogen in years 2, 7 and 11) and L (admixture with lodgepole pine) on a lowland raised bog at Mabie.*

there is continued nitrogen availability. Therefore, on sites of this lithological type, heather control alone will ensure adequate availability of nitrogen.

Data from an experiment at Mabie are presented in Figure 5 to illustrate a category D site type. This is a lowland raised bog with a high percentage of heather in the ground vegetation following ploughing and draining. The treatments presented are herbicide (H), nitrogen application (N) and mixture with lodgepole pine (L) (the latter treatment will be discussed later). Unfortunately, there was no proper control treatment but experiments nearby on the same site type demonstrate that, without treatment, pure Sitka spruce would have declined into severe check by age 10. The herbicide was applied as 2,4-D ester (5 kg a.e. ha^{-1}, or 11 litres product ha^{-1}) in August of the third year and the nitrogen was applied as urea (150 kg N ha^{-1}) in the years 2, 7 and 11. From Figure 5 it is clear that the herbicide had very little long-term effect and the growth rate declined whereas, with regular applications of nitrogen, growth has continued unchecked.

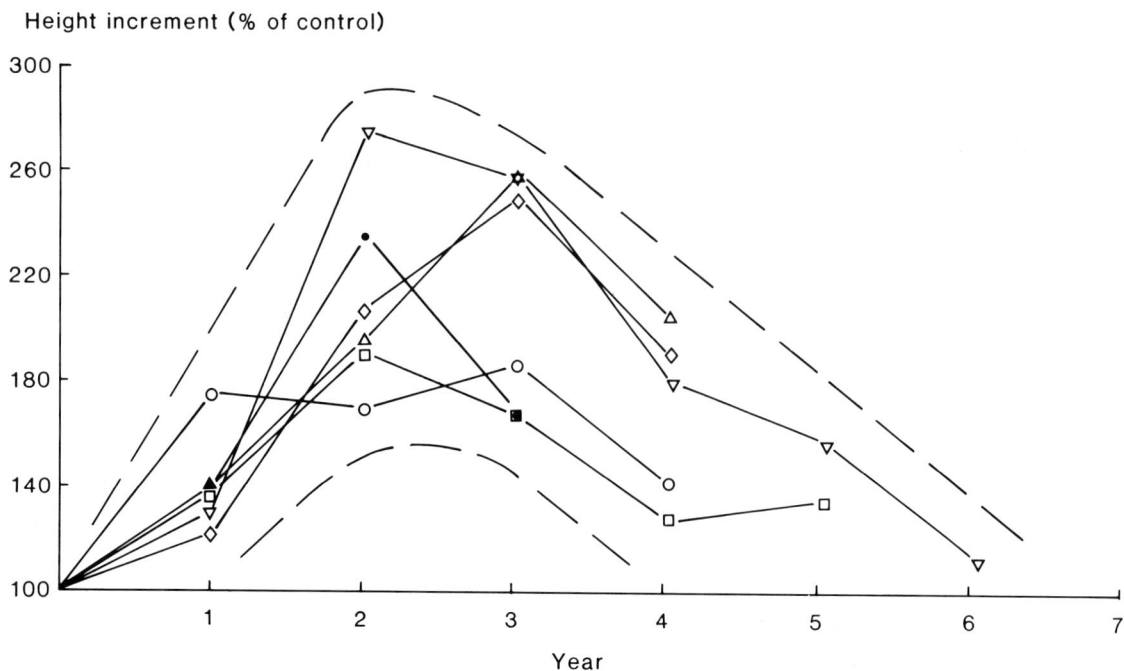

Figure 6. *The annual height increment response to heather control (expressed as a percentage of control) of Sitka spruce at six experiment sites in Scotland.*

Plate 1.
Category A site –
Sitka spruce growing
unchecked on heather/
grass site (brown earth
over Silurian) without
heather control or
nitrogen application.
(C. M. A. Taylor)

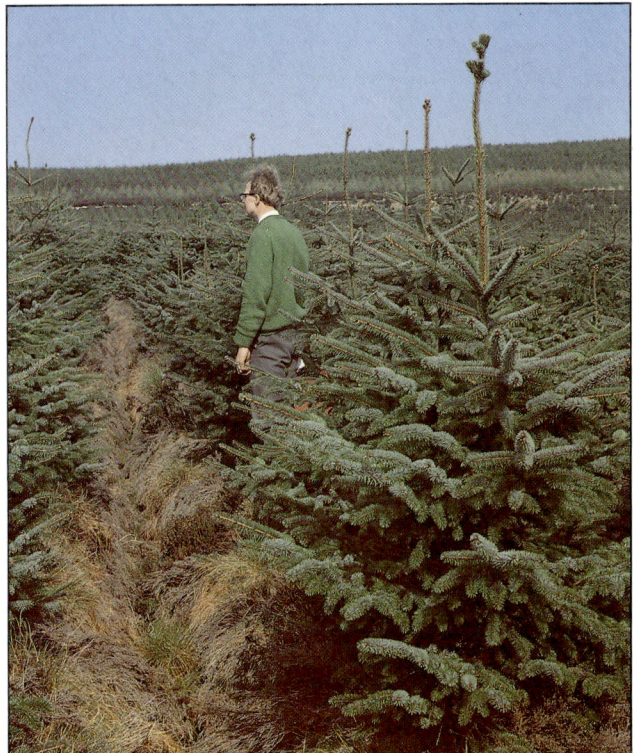

Plate 2.
Category B site –
good growth response
by Sitka spruce to
heather control which
should be sufficient to
enable full canopy
closure (peaty ironpan
soil over Lower Old
Red Sandstone).
(J. M. Mackenzie)

Plate 3.
Category C site –
growth of Sitka spruce
beginning to decline
after response to
heather control on an
ericaceous brown
earth over Ordovician
and will now require
application of nitrogen
to prevent check.
(C. M. A. Taylor)

Plate 4.
Category D site –
Sitka spruce in severe
growth check, despite
previous heather
control, requiring
several applications of
nitrogen to achieve
full canopy closure
(ironpan soil over
Middle Old Red
Sandstone).
(C. P. Quine)

Plate 5.
Sitka spruce crop on a
ploughed restocking
site at Speymouth
(podzolic ironpan over
Middle Old Red
Sandstone, category
D) where growth rate
is beginning to check.
(P. M. Tabbush)

Plate 6.
Right. Pure Sitka spruce in severe growth check due to nitrogen deficiency. *Left.* Same treatment as above, but Sitka spruce growing well due to admixture with Scots pine (1:1). *(C. P. Quine)*

Plate 7.
Sitka spruce on a category D site (ironpan soil over Dalradian quartz mica schist) benefiting from admixture with broom. *(C. P. Quine)*

Category C is a difficult category to demonstrate as it merges into categories B and D depending on the duration of the growth response to heather control. This is indicated in Figure 6 which shows the height growth response obtained from herbicide application relative to the growth of the control for a number of experiments on category C sites. The two outer dotted lines have been added to indicate the 'band of response' which ranges from about 3 to 6 years. Crops on category B sites would follow the same pattern in the first 3 years but the enhanced growth would be maintained, while crops on category D sites would either not respond or the response would be completed within 3 years and growth would relapse.

Treatment decisions for the category C sites are therefore the most difficult due to the variation in the length of the response to a herbicide treatment. The closer the site accords with category D then the more likely it would be that a nitrogen programme would be the best choice. However, it would be unwise to embark on a nitrogen programme if a single herbicide application would have an effect on growth equivalent to two nitrogen applications. In dealing with such sites local experience of growth response to heather control will be the best guide. It is also important to ensure that crop requirements for phosphate and potassium are also satisfied, when considering appropriate treatment for nitrogen deficiency.

The next two sections will try to help the manager categorise sites and give greater detail on the treatments available.

Categorisation of sites

From the experimental evidence and field experience it is clear that nitrogen availability in first rotation plantations is closely linked to soil type (Table 3a) and lithological class (Table 3b), although the relative dominance or frequency of heather is also an important factor.

When identifying the exact category for a particular site the three steps listed below should be followed:

● Step 1
Find the appropriate soil type in Table 3(a)

and commence from the left hand side of the categories listed against that soil type. If only one category is listed then there is no need to continue to steps 2 and 3.
Note: The nitrogen deficiency problems associated with man-made or littoral soils are not covered in this Bulletin.

● Step 2
Identify the appropriate lithology group in Table 3(b). If this lies:
– within group I then move two categories to the right,
– within group II move one category to the right,
– within group III stay in the same category.

● Step 3
If the soil type is mineral or organo-mineral (soil group codes 1, 3, 4, 6 or 7) and the site is dominated by *Calluna* (more than 50% ground cover – equivalent to the 'ericaceous phase' mapped in Forestry Commission soil surveys) move one category to the right. If not, then stay in the same category.
Note: This third step should **not** be applied if the soil is classified as deep peat (i.e. soil group codes 8, 9, 10 or 11).
Note: When recently fire-damaged vegetation is encountered, the dominance rating given to heather will need to be adjusted to that apparent locally on the same site type where there has been enclosure and protection from fire.

Example 1
Inverliever experimental site (see Table 2) – peaty gley soil overlying Dalradian quartzite (more than 50% *Calluna* cover).

Step 1 = category A
Step 2 = category C (group I lithology = 2 categories to the right)
Step 3 = category D (> 50% *Calluna* = 1 further category to the right)

Therefore, Inverliever is a category D site.

Example 2
Strathardle experimental site (see Table 2) – intergrade ironpan soil overlying Dalradian mica schist (more than 50% *Calluna* cover).

Table 3(a). Main soil types of upland Britain categorised by nitrogen availability (soils according to classification by Pyatt, 1982)

Soil group	Code	Soil type	Category
Brown earths	1	Typical brown earth	A
	1d	Basic brown earth	A
	1u	Upland brown earth	A B
	1z	Podzolic brown earth	A B
	1e	Ericaceous brown earth	A B C
Podzols	3	Typical podzol	B C D
	3p	Peaty podzol	B C
Ironpan soils	4b	Intergrade ironpan soil	A B C
	4	Ironpan soil	A B C D
	4z	Podzolic ironpan soil	B C D
	4p	Peaty ironpan soil	A B C
Peaty gleys	6	Peaty gley	A B C D
	6z	Peaty podzolic gley	B C
Surface-water gleys	7	Surface-water gley	A B C
	7b	Brown gley	A
	7z	Podzolic gley	A B C
Basin bogs	8a	*Phragmites* bog	A
	8b	*Juncus articulatus* or *acutiflorus* bog	A
	8c	*Juncus effusus* bog	A
	8d	*Carex* bog	A
Flushed blanket bogs	9a	*Molinia, Myrica, Salix* bog	A
	9b	Tussocky *Molinia* bog; *Molinia, Calluna* bog	A B
	9c	Tussocky *Molinia, Eriophorum vaginatum* bog	B C
	9d	Non-tussocky *Molinia, Eriophorum vaginatum, Trichophorum* bog	B C
	9e	*Trichophorum, Calluna, Eriophorum, Molinia* bog (weakly flushed)	B C D
Sphagnum bogs	10a	Lowland *Sphagnum* bog	D
	10b	Upland *Sphagnum* bog	D
Unflushed blanket bogs	11a	*Calluna* blanket bog	C D
	11b	*Calluna, Eriophorum vaginatum* blanket bog	C D
	11c	*Trichophorum, Calluna* blanket bog	D
	11d	*Eriophorum* blanket bog	D

Step 1 = category A
Step 2 = category B (group II lithology = 1 category to the right)
Step 3 = category C (> 50% *Calluna* 1 further category to the right)

Therefore, Strathardle is a category C site.

On any particular area there is likely to be a complex mixture of site types which may require compromise decisions on treatment. However, by mapping the categories the economic implications of various treatment options are easier to calculate.

Table 3(b). Ranking of the main lithologies according to the likely availability of nitrogen in overlying soils

		Geological map* reference numbers
Group I	*Low nitrogen availability*	
	Torridonian sandstone	61
	Moine quartz-feldspar-granulite, quartzite and granitic gneiss	8, 9, 10, 12
	Cambrian quartzite	62
	Dalradian quartzites	17
	Lewisian gneiss	1
	Quartzose granites and granulites	34 (part only)
	Middle/Upper Old Red Sandstone (Scotland)	77, 78
	Upper Jurassic sandstones and grits	97, 98, 99
	Carboniferous grits and sandstones	81 (part only)
Group II	*Moderate nitrogen availability*	
	Moine mica-schists and semi-pelitic schists	11
	Dalradian quartzose and mica schists, slates and phyllites	18, 19, 20, 21, 23
	Granites (high feldspar, low quartz content)	34 (part only)
	Tertiary basalts	57
	Old Red Sandstone basalts, andesite and tuff	44, 46, 47, 48, 50
	Silurian/Ordovician greywackes, mudstones (Scotland)	70, 71, 72, 73, 74
	Lower and Middle Jurassic sediments	91, 94, 95
Group III	*High nitrogen availability*	
	Gabbro, dolerite, epidiorite and hornblende schist	14, 15, 26, 27, 32, 33, 35
	Lower Old Red Sandstone	75
	New Red Sandstone	85, 89, 90
	Carboniferous shales and basalts	53, 54, 80†, 81 (part only), 82, 83, 84
	Silurian/Ordovician/Devonian shales (Wales and south-west England)	68, 69, 70, 71, 72, 73, 74, 75, 76, 77, 78
	Limestones	24, 67, 80‡, 86
	Cambrian/Precambrian	60, 64, 65, 66

* Reference: Institute of Geological Sciences Geological Map of the United Kingdom (3rd Edition *Solid,* 1979), published by the Ordnance Survey;
† refers to Scotland only;
‡ refers to England and Wales only

Notes
1. Geological Map index no. 34 has been subdivided into: (a) quartzose granites and granulites (Group I), (b) granites with a high feldspar and low quartz content (Group II).
2. Geological Map index no. 81 has been subdivided into: (a) grits and sandstones (Group I), (b) shales (Group III).
3. Where soils occur over drift material, then their characteristics (in terms of nitrogen availability) will be similar to that of the solid parent material from which the drift was derived.

Nitrogen availability on restock sites

Although large-scale restocking of upland forests is still a fairly new practice, particularly on the poorer site types, there is some experimental evidence to indicate whether differences exist in availability of nitrogen between first and second rotation sites.

On the more quartzitic heathland soils there appears to be little difference – spruce will be equally nitrogen deficient in the second rotation as in the first (see Plate 5). This is demonstrated by results from a restock experiment at Speymouth (Figure 2) on a podzolic ironpan soil overlying siliceous Middle Old Red Sandstone where the previous crop had been low yield class Scots pine (YC4). According to the classification for first rotation sites described above, this site would be in category D. Following clearfell and complete ploughing the experiment was planted with Sitka spruce and four treatments have been applied subsequently (Table 4). The initial application of unground rock phosphate at 50 kg P ha^{-1} had no effect on early growth as can be seen by the response depicted in Figure 7. It was only

Table 4. Treatments applied to restock experiment at Speymouth

Year	Treatments			
	O	H	PH	NPH
1	–	–	P	P
6	–	H	H	H
7	–	H	H	H
8	–	–	–	N

Treatment codes: P = application of phosphate at 60 kg P/ha; H = application of 2,4-D herbicide; N = application of nitrogen at 150 kg N/ha.

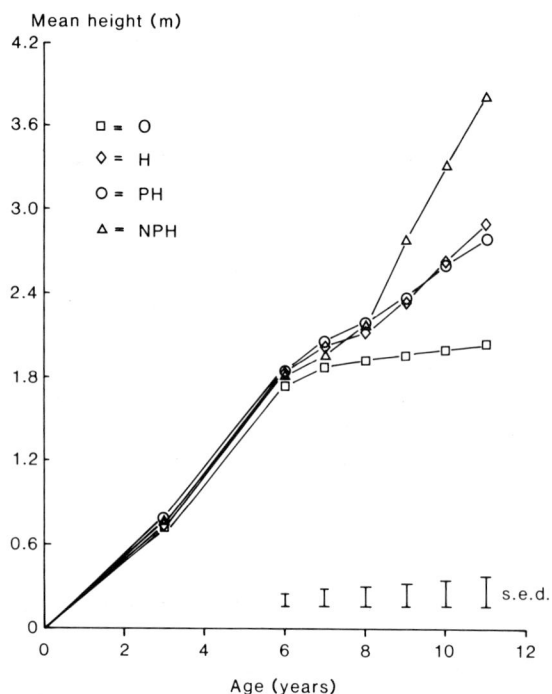

Figure 7. *The effects of treatments O (control), H (heather control in years 6 and 7), PH (H plus application of phosphate in year 1) and NPH (PH plus application of nitrogen in year 8) on the mean height growth of Sitka spruce on a second rotation site (podzolic ironpan) at Speymouth.*

following the herbicide application (2,4-D ester at 6 kg a.e. ha^{-1}, or 13 litres product ha^{-1}) in years 6 and 7 that growth differences began to emerge. The application of nitrogen (urea at 150 kg N ha^{-1}) to treatment 'NPH' in year 8 has resulted in considerably larger growth responses indicating that this site would still be categorised as D in the second rotation.

Similarly, on the peatlands there do not appear to be large differences in nitrogen availability between first and second rotation sites, although phosphate and potassium availability are greatly increased in the early years (Taylor, 1986a). A good example of this is an experiment on a restock site at Inchnacardoch (Figure 2) which had previously carried a lodgepole pine crop (YC8/10). The soil type is an upland *Sphagnum* bog overlying quartzose Moine schists which, according to the classification system, would be a category D site. The experiment was replanted with Sitka spruce and treatments applied (see Table 5) – herbicide

Table 5. Treatments applied to restock experiment at Inchnacardoch and first rotation experiment at Shin.

Year	Treatments			
	Inchnacardoch		Shin	
	H	PKH	H	PKH
1	–	P	–	P
3	–	K	–	K
4	–	–	H	H
6	H	H	–	–

Treatment codes: P = application of phosphate (unground rock phosphate) at 50 kg P/ha; K = application of potassium (muriate of potash) at 100 kg K/ha; H = application of 2,4-D herbicide.

only; or herbicide, phosphate and potassium. By year 10 there were no differences in height growth between the treatments (see Figure 8), which is in marked contrast to what would be expected in the first rotation. For comparison, results from a first rotation site at Shin on an equally poor category D site (weakly flushed blanket bog overlying Moine) with similar treatments (see Table 5) show considerable treatment differences (see Figure 8).

Mean height (m)

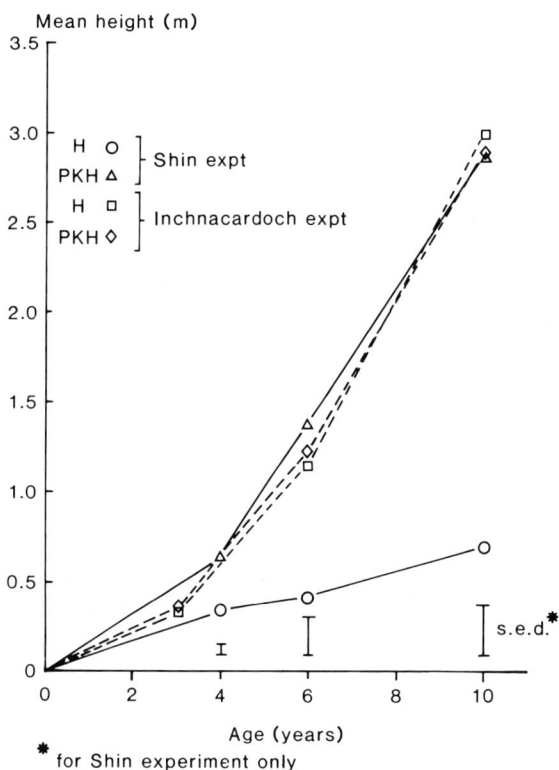

Figure 8. *The effects of treatments H (heather control) and PKH (heather control plus application of phosphate and potassium) on the mean height growth of Sitka spruce on a first rotation site at Shin and a second rotation site at Inchnacardoch, both on deep peat over Moine.*

Foliar N levels (% odw)

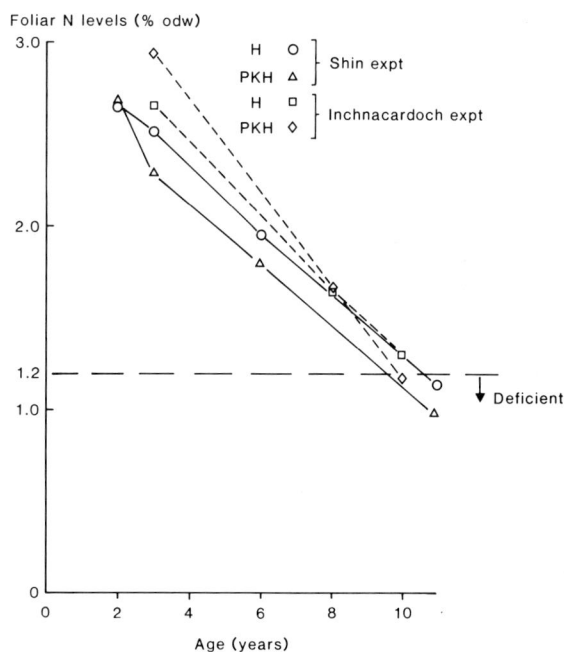

Figure 9. *Decline in foliar nitrogen levels of Sitka spruce for treatments H (heather control) and PKH (heather control plus application of phosphate and potassium) on a first rotation site at Shin and a second rotation site at Inchnacardoch, both on deep peat over Moine.*

The differences between these first and second rotation sites are largely due to the increased availability of phosphate and potassium – treatment H performing as well in the second rotation at Inchnacardoch as treatment PKH in the first rotation at Shin. When the foliar nitrogen levels for both experiments are plotted (Figure 9), it would appear that the pattern of declining foliar nitrogen levels is very similar. Therefore, despite the presumed drying of the peat by the previous tree crop (Pyatt, 1987), there has not been an increase in the rate of mineralisation of nitrogen in the second rotation site. There are, however, indications on other peatland sites that nitrogen deficiency may not be as limiting to growth as in the first rotation but this has not yet been proven experimentally.

At this stage within a given category, there do not appear to be major differences in nitrogen availability between first and second rotation sites for heaths or peatlands. This, however, needs to be fully confirmed by the continued monitoring of these older experiments and the larger number of more recent experiments.

Treatment of nitrogen deficiency

The two main methods of alleviating nitrogen deficiency in a Sitka spruce stand (heather control or application of nitrogen fertiliser) can be applied separately or in combination, depending on the category of the site. This next section presents details on application methods, rates and season of application.

One further method, which has been successful in the past, is also included – rehabilitation of stands by later interplanting of nurse species.

Heather control

Burning is an effective method of heather control prior to cultivation and planting as this considerably lengthens the time before reinvasion by killing all the heather and much of the seed. Alternatively, heather can be sprayed overall with herbicide or mechanically swiped in advance of cultivation. Such treatments are particularly useful on sites with mixed vegetation, when heather can be replaced by grasses (e.g. *Agrostis, Molinia*). Cultivation techniques also influence the onset of heather competition, e.g. shallow ploughing would be more desirable than ripping.

However, it should be borne in mind that on category C and D sites where severe nitrogen deficiency is anticipated, it is advisable to use species mixtures rather than pure Sitka spruce. This is a more prudent prescription for such sites, removing the requirement for substantial expenditure on remedial fertilisation at a later stage.

After planting, the recommended method of heather control is to use herbicides, and up-to-date prescriptions are given in Forestry Commission Field Book 8 *The use of herbicides in the forest* (Williamson and Lane, 1989).

General

Heather is rather difficult to control with herbicide because of its small, waxy leaves and woody nature. It is capable of re-invading rapidly, so spot or band treatments are inappropriate and treatment must be applied over the whole site.

It has proved difficult to control heather on mineral sites with low nitrogen availability, i.e. with increasing difficulty along the site category gradient A to D. In category A the heather is vigorous and presents more foliage area to intercept spray droplets, and the leaf cuticle on young, actively growing shoots is thin and more easily penetrated. Mackenzie (1974) reported that herbicidal effects were enhanced by an application of NPK fertiliser 8-10 weeks prior to spraying, the fertiliser possibly causing a flush of new foliage which allowed enhanced uptake of the herbicide. It is necessary to apply a heavier dose of herbicide on the more difficult sites. It has occasionally proved impossible to achieve satisfactory control on some category D mineral sites despite applying high rates of chemical.

Currently, only two chemicals are recommended – 2,4-D ester and glyphosate (other alternative chemicals are continually under test, e.g. fluoroxpyr and triclopyr are currently being evaluated).

2,4-D ester

The ester formulation of 2,4-D is recommended because it passes more readily through the waxy foliage and bark of heather than water based herbicides (uptake by heather is mainly through its aerial parts). Esters are less easily washed off by rain and can be formulated in special oils for ultra-low volume use without difficulty. The phenoxy herbicides (to which group 2,4-D belongs) are prone to volatilisation in hot weather, and the resulting vapours are very herbicidally active and can drift for long distances. For this reason, only less volatile (iso-octyl) esters of 2,4-D are approved by the Pesticides Safety Division of MAFF for use in forestry.

Timing of treatment

Heather is most susceptible to 2,4-D when it is actively growing. This is generally in the period June to mid-August (similar to the active shoot growth of Sitka spruce), although it starts later and finishes earlier at more northerly latitudes and higher elevations. The season may also be shorter on category D sites than on category A sites. The application season can be extended to include May and late August by increasing the rate of herbicide. Results outside these dates are variable and erratic from year to year.

Rates and methods of application

The approved product for ultra low volume (ULV) application is Silvapron-D, manufactured specially for this use, and containing 400 g l^{-1} of 2,4-D ester. It is applied without dilution through spinning-disc controlled droplet applicators such as the Micron ULVA 16 or Herbi. Recommended rates are given in Table 6.

A number of emulsifiable concentrate formulations containing 500 g l^{-1} of 2,4-D ester are approved for use in forestry and these are listed

Table 6. Application rates for Silvapron-D (2,4-D ester, 400 g l^{-1}). Applied by ULVA (l ha^{-1})

Site category	Mid-July to mid-August	Second half of August
A/B/C/D (peat)	10	12
A/B (mineral)	10	12
C/D (mineral)	15	15

The ULVA should be held low so that a minimum of spray gets on the trees which must be at least 1 metre tall to avoid excessive damage.

If the Herbi is used, the above rates must be increased by 25% to achieve an equivalent effect to that of the ULVA.

in Forestry Commission Field Book 8. They are applied using a semi-pressurised knapsack sprayer, diluted in water at medium volume (200–700 l ha^{-1}) using conventional nozzles, or at low volume (50–200 l ha^{-1}) using very low volume (VLV) nozzles. Earlier applications (May to mid-July) are possible with the ULV formulation because the spray can be directed away from the trees more accurately. Recommended rates are given in Table 7.

Table 7. Application rates for 500 g l^{-1} emulsifiable concentrate formulations of 2,4-D, applied by knapsack sprayer (l ha^{-1})

Site category	May	June to mid-August	Second half of August
A/B/C/D (peat)	10	8	10
A/B (mineral)	10	8	10
C/D (mineral)	12	10	11

Precautions

There is a risk to bees through ingestion if heather is sprayed when in flower, and good liaison with bee-keepers is advisable. Hives should be moved from the vicinity of the treated site, at least until the next heavy rain.

The most important drawback to the use of 2,4-D is its ability to taint water, even at concentrations as low as 1 microgram per litre (0.001 parts per million) after normal water treatment. This restricts the amount that can be applied in a water catchment area at any one time, and water authorities should be consulted when planning the spraying programme.

Glyphosate

Glyphosate is considerably more expensive for heather control than 2,4-D, but has the advantages that it does not taint water and is broken down to harmless residues on contact with the soil. The main disadvantage is that glyphosate is sold as an aqueous formulation which is less rain-fast than 2,4-D ester. It is recommended that at least 6, and preferably 24, hours should elapse between application and significant rain.

Timing of application

Spruce is susceptible to damage during the period of active growth which for Sitka spruce often includes 'lammas' shoots produced in August–October. Little damage results, however, if the spray is carefully directed to avoid unlignified shoots and newly formed foliage. Glyphosate is active on heather throughout the growing season and into early October, extending the control season beyond that possible with 2,4-D. The recommended application dates are from mid-August to the end of September.

Rates and methods of application

Currently, the only approved product is Roundup (360 g l^{-1} glyphosate) and recommended rates are given in Table 8. There is also an additive (Mixture B) which can be used with this product to increase the rate of uptake and ensure effective control, although this also increases the potential damage to Sitka spruce foliage.

The 4 litre rate is almost always adequate on the easier sites, but on the most difficult sites in category D even 6 litres may prove inadequate, and the most appropriate treatment will be to apply nitrogen fertiliser.

Table 8. Application rates for Roundup — 360 g l^{-1} glyphosate (l ha^{-1})

Site category	Mid-August to September
A/B/C/D (peat)	4
A/B (mineral)	4
C/D (mineral)	6

Precautions

No special precautions apply to glyphosate, which has low mammalian toxicity but is classed as harmful to fish. As with all pesticides, operator and environmental safety are controlled through the conditions of approval which appear on the product label. **READ THE LABEL CAREFULLY.**

Application of nitrogen

Type of fertiliser

Urea and ammonium nitrate are the most commonly used nitrogen fertilisers in forestry (Binns, 1975), both being high in nitrogen content (46% and 34% respectively) and water soluble. They give a good compromise between low handling/application costs and fertiliser cost. In terms of growth response there is nothing to choose between them (Taylor, 1987), although urea has tended to be the main choice because it has been cheaper. However, this may not always be the case and relative prices should always be checked before ordering fertiliser.

Although contamination of watercourses is unlikely to occur with conventional forest fertilisation (Taylor, 1986b), urea does have the further advantage that it will not leach as readily as ammonium nitrate. Urea must first be converted to ammonium and then to nitrate, during which process ammonium becomes immobilised in the organic layers (Overrein, 1970). This process is also slowed down in wet, cold soils (Malcolm, 1972) so that, at any given time, only limited amounts of nitrate are available for leaching.

Rate of application

There are two aspects to the rate of fertiliser applied — the frequency of application and the amount of fertiliser applied at each application. The short term nature of the fertiliser response has already been highlighted and McIntosh (1983) estimated that the response period was 3–4 years. Therefore, on very nitrogen deficient sites, 3 to 4 nitrogen applications may be necessary to enable pure Sitka spruce to grow

satisfactorily from the onset of deficiency to full canopy closure, when further application should be unnecessary.

The current standard rate of fertiliser application is 150 kg N ha^{-1} (approximately 330 kg of urea ha^{-1} or 440 kg of ammonium nitrate ha^{-1}). Recent experiments (McIntosh, 1983; Taylor, 1987) suggest that this is a reasonable compromise between rate of fertiliser and growth response obtained. Certainly there would be a significantly lower growth response if the rate was reduced, particularly on the most nitrogen-deficient sites (see Figure 1).

Season of application

Experimental work with urea has indicated that, despite the brevity of the response period, the time of year the fertiliser is applied has no effect on the overall growth response by the crop (Taylor, 1987). However, it would be inadvisable to plan nitrogen fertiliser programmes when there is a risk of run-off over frozen or snow-covered ground. Provided periods subject to these climatic effects are avoided, then there should be no disadvantage from applying nitrogen towards the end of, or even outwith, the growing season compared with the start of the growing season. However, increment in that year will be lost (recouped at the end of the response period) if application is made after June. Late applications (August onwards) should be avoided in upland valleys subject to early frosts.

Application techniques

Three application techniques are available — mechanical, manual and aerial. The first method is unlikely to be much used because vehicle access will be restricted by tree size at the stage when nitrogen fertiliser is required. Manual applications can be very accurate, although distribution of fertiliser and supervision can be time consuming. Care needs to be exercised in the placement of fertiliser and the base of the tree must not be used as a target. High concentrations of nitrogen fertiliser around the root collar can be very damaging and can even lead to tree death (S. C. Gregory, Forestry Commission

Pathology Branch, personal communication). Foliage can be scorched by high concentrations of fertiliser lodging on the needles. The best manual application technique is to broadcast the fertiliser over the whole site, avoiding the furrows, which not only avoids tree damage but is just as effective for crop response.

Aerial application by helicopter is the most common method allowing rapid execution of large programmes and reducing supervisory workload, although it can be prone to erratic distribution dependent on wind speed, terrain, quality of flying and equipment. However, due to the speed of the colour response in nitrogen deficient Sitka spruce the evenness of the distribution can be visually checked a few months after application. (Wait until the following spring in the case of autumn application.) Contracts should include provision for re-application to untreated areas.

Post-planting introduction of nurse species

Successful rehabilitation treatments were carried out in the 1940s in Wester Ross and elsewhere in north Scotland, interplanting severely checked Sitka and Norway spruce crops on moorland sites with lodgepole pine and Japanese larch. Similar operations were successful in checked Sitka spruce plantations on heathlands at Clashindarroch in north-east Scotland. These treatments provide an effective alternative to heather control and/or application of nitrogen.

However, there are several drawbacks to this technique. It could be difficult and costly to successfully establish nurse species due to vulnerability to weed competition and browse, particularly when a considerable period has elapsed since cultivation. The likely cost would have to be compared with the cost of a remedial programme of herbicide and fertilisers. There will also be a delay before the benefit arises (approximately 10 years from time of planting the nurse species). Therefore, it is unlikely that this practice will be widely used, although it should be considered where pure Sitka spruce crops have been very recently planted on sites where severe nitrogen deficiency can be expected and beating-up programmes are planned.

Avoidance of nitrogen deficiency

There are several potential treatments to prevent nitrogen deficiency in Sitka spruce including use of nurse species, nitrogen fixing plants and the application of sludge or lime.

Nursing mixtures

This is the most tried and tested of these alternative treatments, commonly used in the past to nurse Sitka spruce through heather check by planting in mixture with Scots pine, lodgepole pine or larch. It is now recognised that this nursing effect is not just derived from suppression of the heather, but that there is a considerable increase in nitrogen availability (O'Carroll, 1978). As previously indicated, the mechanisms behind this phenomenon are not yet fully understood, but the evidence from experiments and field experience is sufficiently convincing to support the use of this technique in general practice.

One experimental result is included in Figure 5, where the height increment of the L treatment (admixture with lodgepole pine) is equivalent to the N treatment (regular application of nitrogen) following a period of 'semi-check' between years 5 and 9. This latter point highlights the necessity of selecting the correct nurse species, provenance and mixture pattern to prevent the Sitka spruce being suppressed.

Nurse species can be used on both the heaths and the peatlands although species choice and mixture pattern may be different. On the heathlands Scots pine has proved a very satisfactory nurse (see Plate 6), both in the early years and later by allowing rapid dominance by Sitka spruce in the early pole stage. However, choosing planting stock from a suitable seed origin is crucial to ensure good survival and growth to enable the Scots pine to be effective as a nurse. Japanese larch is also a very effective nurse but can be an aggressive competitor in the thicket and pole stage.

On the peatlands the best choice is lodgepole pine which not only nurses the Sitka spruce but is a very effective pioneer on these sites. As with Scots pine the choice of seed origin is important, although in this case it is to prevent early dominance of the spruce by the nurse. Alaskan provenances are the most suitable choice, establishing well but not growing too rapidly in the early years.

Current recommendations are to have a mixture ratio of 1:1, planted in a fairly intimate mixture at normal spacing (Taylor, 1985). In most cases the pattern should be alternate pairs or triplets in every row, planted in a staggered fashion. This ensures proximity of Sitka spruce to the nurse but gives some insurance against suppression, should the nurse prove more vigorous than expected. There is good historical evidence to support row mixtures (alternate single or pairs of rows containing pure planting of Sitka spruce or the nurse), particularly for Scots pine/Sitka spruce mixtures on dry, heathland sites. On sites of low windthrow hazard it is then possible to remove the nurse by line thinning while it is being suppressed. However, there may be some danger in removing the nurse before it has completed its task on the most infertile sites.

The use of nursing mixtures is recommended when afforesting or replanting category D sites, and should also be considered on many category C sites to avoid the need for costly remedial treatment. There are, however, two situations where the use of nursing mixtures would not be successful and pure pine should be planted. These are either where there is a combination of a heathland site over quartzite and low rainfall likely to cause moisture stress in Sitka spruce, or where there is likely to be a high incidence of frost damage to Sitka spruce.

Nitrogen fixing plants

Nitrogen fixing plants are used successfully for nursing in other countries (Davey and Wollum, 1984). In Britain experimentation with leguminous plants, particularly broom (see Plate 7), demonstrated their nursing benefits (Nimmo and Weatherell, 1961). Despite this success the technique was never adopted in commercial practice due to the difficulty of successfully establishing the broom and protecting it from browsing. Therefore it is unlikely to be a dependable alternative to using conifer nurses. Naturally occurring gorse and broom can be

effectively used and should not be removed from crops of Sitka spruce on heathlands. Lupins (*Lupinus* spp.) have not been successful on upland sites although they have been on littoral sands.

Alders (*Alnus* spp.) are another possible nurse and certainly fix considerable quantities of nitrogen (Gessel and Turner, 1974), which would be more than enough to prevent nitrogen deficiency in Sitka spruce. However, again there are difficulties in establishing alders on poor upland sites (particularly peatlands) which militates against their widespread use at this stage (Lines and Brown, 1982).

Sewage sludge

The use of organic wastes containing high levels of nitrogen is another potential treatment on nitrogen deficient sites. Sewage sludge is a readily available and potentially useful source of nitrogen and is currently used on agricultural land. Similarly, effluent from distilleries and breweries has been successfully utilised in the past for reclamation of heathland to agriculturally-valuable grassland. Current research indicates that application of sewage sludge has improved the growth of Sitka spruce on a heathland site in the early years (Bayes *et al.*, 1989) and eliminated the heather, although the duration of response has still to be determined.

Where supplies of sludge are available within 10–20 miles and there is good access to and within the forest, this option has great potential, offering a long-term improvement in the nitrogen availability on heathland sites (Taylor and Moffat, 1989). Advice on suitable sludge sources, techniques and rates of application can be obtained from Forestry Commission Research Division. It is unlikely to be suitable for wet, peaty sites where there would be a risk of run-off.

Application of lime

Applying lime raises the pH of the soil and increases microbial activity, thereby increasing the rate of nitrogen mineralisation in the long term. Experimental results in Britain, however, have not been very encouraging (McIntosh, 1983), although work in Northern Ireland (Dickson, 1977) has indicated long-term benefits in nitrogen availability and tree growth. Lime is costly and difficult to apply at the heavy rates required (5–10 tonnes ground limestone per ha), therefore at this stage this treatment cannot be recommended.

Conclusions

Nitrogen deficiency in young Sitka spruce plantations can be caused by heather competition, low rates of nitrogen mineralisation or a combination of the two. Field recognition of these growth-limiting factors and site categorisation is the first stage in developing remedial treatment programmes. Where heather competition is the major limiting factor, then herbicide control will normally prove to be the most cost-effective treatment. However, as sites become increasingly low in available nitrogen, heather control is less effective and applications of nitrogen fertiliser are more appropriate. This also applies to remote locations, or where there are terrain or climatic difficulties to ground-based herbicide programmes.

On the poorest sites, several applications of nitrogen will be required for satisfactory growth of Sitka spruce and the associated costs are considerable. Alternatively, it is possible to avoid the need for remedial treatment by planting suitable origins of Scots pine, lodgepole pine or Japanese larch with the Sitka spruce in a nursing mixture when afforesting or restocking such sites.

ACKNOWLEDGEMENTS

We would like to thank previous project leaders, J. Atterson, T. C. Booth, J. M. Mackenzie, G. J. Mayhead and R. McIntosh, who initiated the experiments reported in this paper and the various research foresters who have maintained and measured them over the years. Thanks are also due to D. G. Pyatt and D. B. Paterson, who have made numerous helpful comments on the text and content, and I. M. S. White for the statistics. We are grateful to Site Studies (South) Branch for undertaking the foliar analyses.

REFERENCES

BAYES, C. D., TAYLOR, C. M. A. and MOFFAT, A. J. (1989). Sewage sludge utilisation in forestry: the UK research programme. In, *Alternative uses of sewage sludge,* Conference proceedings, University of York, Sept. 1989. WRC, Medmenham, UK.

BINNS, W. O. (1975). *Fertilisers in the forest: a guide to materials.* Forestry Commission Leaflet 63. HMSO, London.

BINNS, W. O., MAYHEAD, G. J. and MACKENZIE, J. M. (1980). *Nutrient deficiencies of conifers in British forests.* Forestry Commission Leaflet 76. HMSO, London.

DAVEY, C. B. and WOLLUM, A. G. (1984). Nitrogen fixation systems in forest plantations. In, *Nutrition of plantation forests* (eds G. D. Bowen and E. K. S. Nambiar), 361–377. Academic Press, London.

DICKSON, D. A. (1977). Nutrition of Sitka spruce on peat – problems and speculations. *Irish Forestry* **34**, 31–39.

DICKSON, D. A. and SAVILL, P. S. (1974). Early growths of *Picea sitchensis* (Bong.) Carr. on deep oligotrophic peat in Northern Ireland. *Forestry* **47**, 57–88.

EVERARD, J. E. (1973). Foliar analysis sampling methods, interpretation and application of the results. *Quarterly Journal of Forestry* **67**(1), 51–66.

FRASER, G. K. (1933). *Studies of certain Scottish moorlands in relation to tree growth.* Forestry Commission Bulletin 15. HMSO, London.

GESSEL, S. P. and TURNER, J. (1974). Litter production by red alder in western Washington. *Forest Science* **20**, 325–330.

HANDLEY, W. R. C. (1963). *Mycorrhizal associations and Calluna heathland afforestation.* Forestry Commission Bulletin 23. HMSO, London.

LAING, E. V. (1932). *Studies on tree roots.* Forestry Commission Bulletin 13. HMSO, London.

LEYTON, L. (1954). *The growth and nutrition of spruce and pine in heathland plantations.* Imperial Forestry Institute Paper No. 31. IFI, Oxford.

LEYTON, L. (1955). The influence of artificial shading of the ground vegetation on the nutrition and growth of Sitka spruce in heathland plantation. *Forestry* **28**, 1–6.

LINES, R. and BROWN, I. (1982). Broadleaves for the uplands. In, *Broadleaves in Britain* (eds D. C. Malcolm, J. Evans and P. N. Edwards), 141–149. Institute of Chartered Foresters, Edinburgh.

McINTOSH, R. (1983). Nitrogen deficiency in establishment phase Sitka spruce in upland Britain. *Scottish Forestry* **37**, 185–193.

McINTOSH, R. (1984). *Fertiliser experiments in established conifer stands.* Forestry Commission Forest Record 127. HMSO, London.

MACKENZIE, J. M. (1974). Fertiliser/herbicide trials on Sitka spruce in east Scotland. *Scottish Forestry* **28**, 211–221.

MALCOLM, D. C. (1972). The effect of repeated urea applications on some properties of drained peat. *Proceedings 4th International Peat Congress Finland* **3**, 451–460.

MALCOLM, D. C. (1975). The influence of heather on silvicultural practice – an appraisal. *Scottish Forestry* **29**, 14–24.

MILLER, H. G. (1981). Forest fertilisation: some guiding concepts. *Forestry* **54**, 157–167.

MILLER, H. G., ALEXANDER, C., COOPER, J., KEENLEYSIDE, J., MACKAY, H., MILLER, J. D. and WILLIAMS, B. L. (1986). *Maintenance and enhancement of forest productivity through manipulation of the nitrogen cycle.* Results of a study carried out under the European R&D programme in the field of wood as a raw material (1982–85). Final report to the CEC on Contract No. BOS/093 UK.

NIMMO, M. and WEATHERELL, J. (1961). Experiences with leguminous nurses in forestry. *Forestry Commission Report on Forest Research 1961,* 126–147.

O'CARROLL, N. (1978). The nursing of Sitka spruce 1. Japanese larch. *Irish Forestry* **35**, 60–65.

OVERREIN, L. N. (1970). Immobilisation and mineralisation of tracer nitrogen in forest raw humus. *Plant and Soil* **32**(1), 207–220.

PYATT, D. G. (1982). *Soil classification.* Forestry Commission Research Information Note 68/82/SSN. Forestry Commission, Edinburgh.

PYATT, D. G. (1987). Afforestation of blanket peatland – soil effects. *Forestry and British Timber* **16**(3), 15–17.

READ, D. J. (1984). Interactions between ericaceous plants and their competitors with special reference to soil toxicity. *Aspects of Applied Biology* **5**, 195–209.

TAYLOR, C. M. A. (1985). The return of nursing mixtures. *Forestry and British Timber* **14**(5), 18–19.

TAYLOR, C. M. A. (1986a). Upland restocking – nutrition prospects. *Timber Grower* 101, 23.

TAYLOR, C. M.A. (1986b). *Forest fertilisation in Great Britain.* Proceedings No. 251, The Fertiliser Society, London. (23 pp.)

TAYLOR, C. M. A. (1987). The effects of nitrogen fertiliser at different rates and times of application on the growth of Sitka spruce in upland Britain. *Forestry* **60**(1), 87–99.

TAYLOR, C. M. A. and MOFFAT, A. J. (1989). The potential for utilising sewage sludge in forestry in Great Britain. In, *Alternative uses of sewage sludge,* Conference proceedings, University of York, Sept. 1989. WRC, Medmenham, UK.

WEATHERELL, J. (1953). Checking of forest trees by heather. *Forestry* **26,** 37–40.

WILLIAMSON, D. R. and LANE, P. B. (1989). *The use of herbicides in the forest.* Forestry Commission Field Book 8. HMSO, London.

ZEHETMAYR, J. W. L. (1954). *Experiments in tree planting on peat.* Forestry Commission Bulletin 22. HMSO, London.

ZEHETMAYR, J. W. L. (1960). *Afforestation of upland heaths.* Forestry Commission Bulletin 32. HMSO, London.

Printed in the United Kingdom for HMSO
Dd 291283 C25 3/90